The Incalculable Essays

Miguel A. Sanchez-Rey

Table of Contents

The Sun Never Sets

5-8

The Scientific Age and the Scientific Process

10-15

Response to Wesley Nolan Mattingly's, "The Ecology of Being"

17-19

PHPR Protocol

21

The Advance Age

23-28

The Sun Never Sets

[Author: Miguel A. Sanchez-Rey]

The sun never sets is the motto of a far away nation-state at the dawn of the Scientific Age. It's scattered across different continents and it's besieged by a neo-colonial nation whom decline has been ongoing since the end of the Second World War. A neo-colonial nation that prides itself as a nation of immigrants, that values itself in it's freedom of speech, and revels in it's tolerance of race and ideological opinion. Yet its economic and military policies have revealed the U.S. to be far-right in comparison to other developed countries that has since then refused to partake in the dominance of free-markets, the rise of a service base economy, and the willingness to reward private interest while slashing domestic spending and encouraging the radicalization of the political spectrum.

It's an undiscovered country that has become a nation of cult-leadership and flaw policy making. A country that has gone its separate ways and establish itself as a model country of savage cuts to public spending, poor health-care and stagnate wages. It has failed to live up to standards when it comes to education as the European states or even measure up to the infrastructure of Hong Kong. It has since then revealed itself to be nation-state on the verge of authoritarianism as war-like and free-market policies take over the American landscape. It has revealed a deeply divided and paranoid nation that expresses fundamental religious divisions and class indifferences. Only then has the undiscovered country been besieged by enormous anger and hate with the political establishment profoundly divided.

The U.S. has become a cult. A cult-following has ensued across different party lines. Where leadership must follow orders from the private sector. But let's be exact. The U.S. was, in many different respects, bound to detach itself from the constitutional order that laid the groundwork for the Federalist policies of the early American founding figures. Since then the constitutional order has become more inert. With each accumulating decade the U.S. Constitution became more irrelevant to social progress. Social progress that is both cognizant of international necessity of human and economic rights. The U.S. nevertheless has become victim to private markets. Where rewarding the private sector became the U.S. national norm. Compromising the achievements of the 1960's New Left and the civil rights era.

Other countries have pursued policies which are more open to social progress; that though seemingly populace, has favor public social spending to pacify the civilian population. But as well has remained independently self-governing but detach from the demands of the American leadership.

The far-away nation, in which the sun never sets, has seen the transition from classical imperialism to a scattered nation across different continents. A nation independently self-governing but united in its core ideal of a common interest. A common interest that can be said to be the internationalist model. The internationalist model that is antagonist to suppression and domination but more incline to pursue mutual economic trade and through its military alliance across the different continents they shield themselves from any external threat to its independence and longevity. That is, the sun never sets on the Common Wealth of Nations.

Only then has American society seen itself more beleaguered by national strife. It has seen itself more on the verge of being divided states with separate economic motives that is detach from the federalist commitment to the House of Congress and the Executive Office. It has use its militaristic policies to implement a suppressive military industrial complex that wagers battle grounds and animosity. The U.S. being a surrounded nation, has since then witness the setting of the sun across the different regional states.

Only time will tell when the U.S. succumbs to third world policies. Where much of the planet, besides the U.S., pursues the internationalist norm of self-governance and a military alliance that stresses common economic interest.

The U.S. is a defiant nation but also fooled by a flaw constitutional order. A constitutional order that has not been able to adapt to modernization and social progress. Subjecting the U.S. to unreasonable policy decision making while its war-like policies has crippled the U.S. capacity to sustain a vibrant military alliance. The disenchantment with the United States growing protectionist movement has rendered the U.S. incapable to manage world-affairs. U.S. decline becomes imminent and ominous. With national animosity more apparent to the United States economic and political establishment.

The Scientific Age and the Scientific Process

[Author: Miguel A. Sanchez-Rey]

Planetary society has undergone false modernization. With progress in industrialization and the high-tech industry economic growth has continued to rise but poverty has rising world-wide and a world that what once thought to be undergoing a peaceful transition has, nevertheless, seen the rising force of political instability. A civilization gripped by the ghostly apparition of Post-Modernity.

That what was seen as an attempt to bridge the differences between the warring factions of the fascist and the social justice movement has morphed into a neo-fascist alliance.

Questioning the legitimacy of the welfare state. That is the experimentation of the state has taking center stage in the preoccupation of political economy. The crumbling of legitimate state federalism into inadequate state sovereignty.

A world in psychosis, brought on by the breakdown cause in the 2016 Presidential elections in the United States, has made the constitutional order insufficient. Democratic principles are now transitioning into an authoritarian stipulation. An authoritarian stipulation where the mass media is disenfranchised and the internet age becomes more regularized as to pacify the population from a cult phenomenon of bad-decision making that causes havoc to the ruling power-structures.

The state has become an inactive organism. An inactive organism that is both docile and complacent. Docile in its tolerance of popular policy and complacent in its acceptance of neo-fascism. That is in itself requires that the military industrial complex absorb the state to regain active function and to shut the general public from the scientific process, and the affairs of state, to relinquish any further revolt against what remains of the benevolence of the marginalized state.

That the British state is to be dismantled and what remains is to be integrated with the European Union, the Common Wealth of Nations is to be wildly strong in its independence, the U.S. federalist institution is to transition into a scientific dictatorship, the People's Republic of China is to continue to slowly develop, and that much of the Middle East and Africa is to remain regionally independent with rising economic potential and increasing exploitation. Lasting damage has been done to the religious state and now planetary society must move forward into matters which are more formal and far-reaching.

The scientific order has shifted. With the triumph of the gravitational wave experiment the scientific process has been put into chaos. The political economy, unable to be burden with the instability of scientific progress, desires to detach the general public from the decisions of the scientific process to temper the Scientific Age.

With the shocking let down at CERN, with the non-realization of utopia at ITER, and then, the chaos brought on by the LEGOS observatory, one has reach the Scientific Age in the form of an uncontrollable scientific machine.

Which its resolution demands that the general public be periodically shut-out from the scientific process. But because the general public desires, nevertheless, to participate in all things which a democratic society normally constitutes, the only avenue in which to quell a revolt against the Scientific Age is to enact authoritarian control of the public sector and to unlink the scientific process from open scrutiny.

Matters which were once accessible to public decision-making in the affairs of the sciences must be temporally dismantled to protect the civil order. Either/or an insane planet, in which persistent psychosis becomes tantamount, leads to an ecological and environmental catastrophe at the finality of human civilization. Whereby much is short-live.

There is in all unanimity against the authoritarianism of the scientific process but such unanimity must go through pacification. The scientific process, being in relation to the authoritarian regime of a scientific dictatorship, cannot see itself as co-existing side by side the democratic movement. For which the only option is to put planetary society into a serene psychological state.

Much is not to change in civil society until the planetary system gradually recovers from collective psychosis. It's to remain in such a state, even as things quiet down, as technological innovation gradually progresses. The public remains apart from political economy and scientific decision-making. Living their lives quietly and accordingly until the scientific process is ready to be disclosed. The world, nevertheless, will seem not out of the ordinary but transitioning into space-exploration in hopes of a far-reaching political society that is self-ruling and a political economy in which trade and market principles is obligatory to a world-wide command economy.

The Scientific Age has revealed a world that is haunted by the apparition of Post-Modernity. An apparition that lurks in the collective subconscious. A society that overreaches itself into matters of statehood encouraging the radicalization of the sciences. That flaw decision-making, at the dawn of the Scientific Age, led to widespread havoc. And that in which cry havoc has let loose mass opposition to the state and in which cry havoc instigated the general public to exceed itself in the scientific process. And in which its resolution requires that the mainstream population make the sacrifice to relinquish their rights to open scientific participation so that no further harm is done to them or the scientific process.

That the political economy be democratically self-manage while the central authority governs with authoritarian sovereignty, to protect the self-management of democratic political economy, until it is ready to be dismantled. At that point the scientific machine becomes manageable to open participation.

The Scientific Age and the scientific process is a delicate distinction. And to reconcile their distinction is to see a planet quietly awaiting the results of an interplay until the interplay completes its last task.

Response to Wesley Nolan Mattingly's, "The Ecology of Being"

[Author: Miguel A. Sanchez-Rey]

Mr. Mattingly's eco-phenomenology of the sense; that in which, the history of ontology is grounded on the untold history of experience, is nevertheless a nonsensical historical account of the lived experience in which the sense is misconstrued as a lived-ontology of historical naturalism. Though Mr. Mattingly presents an eloquent treatment of the phenomenological dynamics of the experiential mechanism it does not suffice that sense, the medium of perceptual acuity, can be use to unearth the wild being of the lived experience. Wild being, conjectured by the early phenomenologist, is an ineffective answer to the catastrophe of ecology. In which wild being is a denotation of the undisclosable otherness.

The untold history of experience is an unintelligible phenomenological reduction. For the untold is yet to be said and the untold is yet to be understood. And that in which experience becomes the unsaid the lived experience cannot account for experiential consciousness that has historical manifestation to *logos*. For logo-centrism is a necessary component in which genuine experiential knowledge is ascertain.

Genuine experientialism cannot be said to be genuine without a foundationalist account of the sciences of perceptual experience. For which without the science of perceptual experience, that in which the sciences seek to ground experientialism, the lived experience cannot be acknowledge as a corporeal being. Without the sciences the corporeal being is lost in a wilderness of nowhere in which the transcendental will not be able to direct the corporeal beings' predicament of historical dichotomy.

The eco-phenomenology of the sense becomes unexplainable to the historical consciousness. In which there is no historical justification for the ontological of the untold. And that in which there can be no cross-fertilization of being-in-the world and being-of-the-earth. For which being-of-the-earth is a being that is ontologically limited and in which being-of-the-earth is trivial to cosmological dogmatism. In which catastrophe ensues. And that in which the history of experience becomes the unsaid and the ecology of being becomes the untold.

PHPR Protocol

Government is a resolution to the state-of-nature. The Physicalist Program [PHPR] is design as a resolution to a foreseeable catastrophic scenario in the Scientific Age in the form of a task. The Grandmaster is to complete a task and set the next task.

The First Task is a 100 Year Task.

PHPR top-scientists are task to complete The First Task.

PHPR top-scientists are selected into PHPR.

Selection of top-scientists into PHPR is a stringent task.

PHPR top-scientists exemplify combat leadership and excellence in scientific scholarship.

Identity of PHPR top-scientists are to remain classified until their task in The First Task is partially declassified.

By then top-scientists are to retire from PHPR.

Acknowledging PHPR top-scientist as The Master of Space-Time.

The Grandmaster will continue on…

The Advance Age

[Author: Miguel A. Sanchez-Rey]

[An Incalculable Age]

The Advance Age is an incalculable age. An age that is wildly strong in its independence and prideful in its civility. An age in which the boundaries of sexuality and the conception of the deity begin to breakdown. Not only does the Advance Age deal with matters which cannot be openly discuss but also it deals with themes that are unheard of.

The Advance Age is an age of wild strength. One in which the futility of being is tantamount and in which all things are bound to be forgotten. All things about the post-modernist era is in, itself, at the eve of the Advance Age a forgotten era of trivialities and non-sequiturs.

The scientific state reigns dominant. The population of planet Earth lies scatter throughout the solar system in space-habitats traveling in an out of the solar habitat on a daily Earth bases to ascertain the far reaches of Milky Way. Not only to discover new worlds that are suitable for mining and mineral extraction, not only to find new energy resources, or to lay the course for uncharted territory, but for sake of humanity to build new bridges that will bring about an internationalist model throughout the Goldilocks zone.

One in which humanity must find the will-power to move beyond its habitable niche and settle elsewhere by using the terraformic process. It's then that at the Advance Age that sexuality becomes a sacred act and that in which in its sacredness a rare if not uncommon act.

At the dawn of the Advance Age the scientific state is confronted by the sacrifices that were made at the eve of the Scientific Age. A world that at the time fell into collective psychosis must shut-out the general public from the scientific process to temper an uncontrollable scientific machine; a wild child, that so many abandoned supposing that the wild child is left to care for itself while it runs amuck. Only then by taking in this wild child in and tempering her can her psychosis be alleviated. And as such can the wild child find independence and strength that becomes a wildly strong predisposition. A wildly strong independence that is manageable to others and open to participation.

The scientific state, prideful in it's power-structure and strength, must confront the realization of the last task. That in which one's privacy becomes more infringe upon and in which collective consciousness begins to tear apart the private subconscious processes of the mind that horrid acts of sexual mutilation become a lurking conundrum. There the council of the scientific state, celebratory of its sexual power and longevity, becomes confrontational with the self-management of democratic political economy. And in which the scientific state, ever more bigoted, ignorant of the scientific process, and broken, realizes the inevitable outcome.

To protect existentiality from violent revolt the scientific state is dismantled. Finding wild strength, in the form of strong-independence, one regains the private sex act. With the self-giving capacity to anticipate does Anarcho-syndicalism become fruitful and long-lasting. Where the boundaries of sexuality and the conception of the deity, in the incalculable age, breakdown.